THE SCIENCE OF CRYSTALS

THE MCGRAW-HILL HORIZONS OF SCIENCE SERIES

Universal Constants in Physics, Gilles Cohen-Tannoudji

The Dawn of Meaning, Boris Cyrulnik

The Realm of Molecules, Raymond Daudel

The Power of Mathematics, Moshé Flato

The Gene Civilization, François Gros

Life in the Universe, Jean Heidmann

Our Changing Climate, Robert Kandel

The Language of the Cell, Claude Kordon

The Chemistry of Life, Martin Olomucki

The Future of the Sun, Jean-Claude Pecker

How the Brain Evolved, Alain Prochiantz

Our Expanding Universe, Evry Schatzman

Earthquake Prediction, Haroun Tazieff

FRANÇOISE

BALIBAR

THE SCIENCE OF CRYSTALS

McGraw-Hill, Inc.

New York St. Louis San Francisco Auckland Bogotá
Caracas Lisbon London Madrid Mexico
Milan Montreal New Delhi Paris
San Juan São Paulo Singapore
Sydney Tokyo Toronto

English Language Edition

Translated by Nicholas Hartmann
in collaboration with
The Language Service, Inc.
Poughkeepsie, New York

Typography by AB Typesetting
Poughkeepsie, New York

Library of Congress Cataloging-in-Publication Data

Balibar, Françoise.
[*La Science du cristal*. English]
The Science of Crystals/Françoise Balibar.
104 p. cm. —(The McGraw-Hill *Horizons of Science* series)
Translation of: *La Science du cristal*.
Includes bibliographical references.
ISBN 0-07-004449-X
1. Crystallography. I. Title. II. Series.
QD905.2.B3413 1993
548—dc20 92-19607

The original French language edition of this book
was published as *La Science du cristal*, copyright © 1991,
Hachette, Paris, France.
Questions de Science series
Series editor, Dominique Lecourt

This book is printed on recycled, acid-free paper containing a minimum of 50% recycled de-inked fiber.

TABLE OF CONTENTS

INTRODUCTION

Here is the crystal in its full scientific glory. Well-entrenched for several years at the heart of basic research in physics, the science of crystals is now expanding into chemistry, the life sciences, astrophysics, and other disciplines. The scientific applications of crystals are part and parcel of the most spectacular breakthroughs in today's technology, in electronics as well as in medicine and pharmacology. Françoise Balibar paints an enthralling picture of these advances.

But in what may be viewed as a historical irony, crystals are at the same time gaining notoriety in a very different arena. As heirs to a long history fraught with legend and mystery, they are once again attracting "healers" who practice acupuncture and acupressure, Ayurvedic medicine, "aura rebalancing," radionics, magnetism, naturopathy, sound therapy, color therapy, etc. Crystals are spawning far-reaching magical views of the world, highly syncretistic in tone, which reconcile Oriental doctrines, Christianity, and Judaism. As such, they have become an ancillary to prayer in modern cosmic religions. An interesting sign of the times is that while the community of crystallographers remains deliberately and often aggressively ignorant of the "gurus" who bustle around the object of their research, those same gurus, in contrast, are always on the lookout for any theoretical element borrowed from basic research which might confirm their convictions and their teachings.

Strenuous efforts are being made to demonstrate that quantum theory itself has established the validity of ancestral practices and intuitions. The curative properties of crystals are now being attributed to "cosmic fluxes" of "subtle energies," and to infinitesimal "vibrations" that resonate through them and indeed through our own bodies. Everything is apparently a matter of resonance. To reassure its adepts and gain the confidence of its clientele with endless courses and seminars, modern magic cannot survive without scientific authority: it presents itself not as an aggressive anti-science, but modestly as a "parascientific" doctrine. "*Para*," translating from the Greek, means "alongside," certainly, perhaps even "marginal," but it implies nothing derogatory. Quite the contrary, suggest the devotees: it is sort of a complement. Those on the other side might call it vice paying homage to virtue.

This concomitant dual triumph, both sacred and profane, is the culmination of two histories. One—brief but eventful—is that of crystallography since the end of the 18th century; the other is immemorially old, and its shadows have not been swept away by the dawn of a rational awareness of crystals. Crystals are seen as perfect Bachelardian objects, a veritable hotbed of "epistemological obstacles" that have arisen from the depths of thought itself to hold back its investigative efforts. As we shall see, this is why it is in the very nature of crystallography to be continually in search of itself. From the 18th century to the present day, this discipline seems destined for a sort of errancy or epistemological nomadism: should we regard it as a chapter in mineralogy (but which one)? Is it part of chemistry? Or perhaps of physics? Hence the fact, at first sight odd, that the scientific horizon addressed here is the science of crystals itself, to the extent that it *is* a science.

According to archaeologists, rock crystals (quartz) have been discovered in the dwellings of the ancestors of the Yuma Indians in California, whose cultural remains date back 8,000 years. Along with gems and precious stones used for jewelry and the goldsmith's and silversmith's trade, they are found at archaeological sites in both North and South America. It appears that quartz crystals were part of the traditional medical armamentarium of the shaman. It is also known that particular properties were ascribed to crystals in the great texts of ancient India, the *Atharvaveda* and the *Ayurveda*. Similar properties and uses have also been found in the Taoist tradition of China.

The familiar fortune-teller's practice of predicting the future with a crystal ball thus has its roots in a long and fascinating past. Because of its perfect shape, and the clean geometry of its facets and their angles, rock crystal is thought to have a direct rapport with the cosmos: an open door to other times and other places. Was it not said that Atlantis supplied itself with energy from gigantic crystals?

The only way to build a rational understanding of crystals is to sweep away these legends. But the question of the origin of rock crystals will always haunt the human imagination; that preoccupation may indeed exclude any observation of the crystal itself by focusing on the perfection of its external form. For a long time, from the 14th to the end of the 18th century, crystals were never themselves the object of research or even of precise observations; instead they were mere arguments used in favor of various cosmogonies. They excited fantastical ideas about the structure of the world and the history of the Earth.

Jean Buridan (1300–1358), several times rector of the University of Paris from 1328 on, and a powerful nominalist logician and highly influential philosopher, explained the formation and fate of the Earth in his *Questions* on Aristotle's *Meteorologica*. Beginning with the phenomenon of erosion, which he himself had studied in the Cévennes region of

France, he postulated a twofold process to explain the future of our planet: a continual uplifting of rocks from its center (this being the origin, in particular, of mountain-building or "orogenesis"); and an erosive phenomenon which destroyed the highest elevations, thus allowing those portions of the Earth located at its center to emerge. This explained the presence of rock crystals at the surface: their "coagulation," he wrote, was due to the extremely intense cold that they experienced for prolonged periods in the bowels of the Earth, before they were transported and deposited into one of the three crusts which cover the planet.

This Aristotelian idea of the formation of crystals by the "coagulation" or "congelation" of water would be transmitted, in a number of scenarios, down to the end of the 17th century. The spectacular stalactite formations found in caves and grottoes would serve as references and models to explain every aspect of the formation of the stones, crystals, and diamonds found in Nature.

The great Johannes Kepler (1571–1630), musing in his charming *Strena seu de nive sexangula* (1611) on the hexagonal shapes of snowflakes and crystals, uses Aristotle's arguments as a basis for stating their origin as liquid, and, although he did compare their internal structure to that of a honeycomb—small spherical elements stacked so as to occupy as little space as possible—explains the arrangement of their faces by invoking reasons drawn from the wellsprings of his mystic cosmology, stating that the geometric forces of crystals are the reflection of the "Earth's soul." We see here Kepler's "Pythagoreanism," the very conviction on which he based his formulation of the laws of gravity which held the planets in orbit around the Sun (however shocking this speculative belief might appear to positivist minds).

The first thinker to reject, on a reasoned basis, the idea that crystals were produced from water was undoubtedly the French chemist Étienne-François Geoffroy (1672–1731). "The simplest stone of all, the most homogeneous, and thus the most perfect, is rock crystal." But it is not frozen water, he states, and then immediately launches another idea with a great future ahead of it, that of a "crystalline essence" or "subtle fluid" as the agent of crystallization. René-Antoine Ferchault de Réaumur (1683–1757) echoed this thinking a short time later.

So we find that although there were at that point divergent ideas about modalities and processes, the link between "crystallization" and cosmogony still had not been undone. One might even say that this link became closer and closer during the 17th and 18th centuries.

The Swiss naturalist Louis Bourguet (1678–1742), for example, invoked the Cartesian theory of the formation of the Earth to uncover "the plan used by Nature in creating crystals." On the other hand, Georges-Louis Leclerc, Comte de Buffon (1707–1788) integrated it into his strictly Newtonian view of Nature and in general, saw in the "depiction" of minerals a rudimentary outline of "organization," a "rough draft of life" due to the presence of "organic molecules" deriving from animal and plant remains deposited in the limestone. "Pyrite," he wrote, "is a mineral of regular form which could not have existed before the birth of animals and plants."

Buffon had the opportunity to use the first observations made under the microscope, an instrument invented, in the true sense of the word, in 1673 by the Dutch scientist Anton van Leeuwenhoek (1632–1723), who studied sea salt, sugar, sand, and rock crystal, and discovered tiny particles in them. An entire generation would then try to correlate these microscopic observations with the law of universal attraction in a vain attempt to explain the genesis of crystals.

The situation reversed itself with Jean-Claude de Lamétherie (1743–1817), the indefatigable editor and proprietor of the *Journal de physique*: now it was crystallography itself which absorbed cosmogony and all the other sciences with it. "Crystallization," he wrote, "must be considered the fundamental principle of the greatest phenomena of nature," and even "the only universal phenomenon." He then gives examples that fire the imagination: "The Earth itself, like the minerals which cover it, formed from a fluid which crystallized as it cooled." But also: "I regard reproduction as a kind of crystallization," hence the pages describing the "crystallization of the fetus!" And finally the soul itself, seat of all feelings and thoughts, emerges as the fruit of the crystallization of the being whose center it occupies.

With Buffon and even more so with Lamétherie, we are just skirting the edges of another thesis—but in this case one present as an element integrated into their general system of interpreting Nature—which, highly developed in its own right, constituted a major and stubborn obstacle to the development of a true science of crystals: the idea of crystallization as an organic process.

The famous French botanist Joseph Pitton de Tournefort (1656–1708) left us a grandiose version of this view of crystals. In his eyes, all organic life could only be of vegetative origin; as Hélène Metzger, the late 19th–20th century historian of science, has written, Tournefort "sees botany everywhere." Corals and sponges, according to him, are not animals, but plants; so why not other minerals as well? "Since there are rock-like plants, it is a presumption admissible in physics that rocks could be plants," runs one delightful passage. The proof: "There can be no doubt that organized stones exist."

And now the word "germ"—which, as we shall see in the description of crystals, was to be used in other, more expert ways—made its entrance: "The various species of pyrites, rock crystals, and a multitude of other rocks are just as suggestive of particular germs as ordinary mushrooms, truffles, and several species of moss, whose seeds have not yet been discovered."

Considering the "assumed form" of crystals, Tournefort refused to see in it the hand of an intelligent Architect; instead, he imagined that they were composed of "kinds of eggs," similar to the seeds "of plants, fish, and birds." The idea was not new—the Renaissance swarmed with these kinds of discourses and conjectures about the origin of rocks—but never before had the idea been expanded so systematically, and moreover in the context of a work which made a name for itself in the history of botany.

Given this evidence, must we then conclude that until the end of the 18th century, crystals were merely the object of unbridled speculation? Many observers did in fact contribute to an understanding of certain properties of crystals, but Dr. Balibar rightly points out that most of these observations, although liberated from the shackles of the great cosmogonic systems, served some purpose other than an understanding of crystals *per se*. The matter at hand was essentially the vexed question of the nature of light; and crystals were involved solely as instruments in these investigations, which were widespread in the 17th century. Some of the most important physical characteristics of crystals were indeed identified along the way, but only in order to resolve conflicts of interpretation relating to optics. The discovery in 1672 of the double refraction of Iceland spar, made by the Danish physician and physicist Erasmus Bartholin (1625–1698), is nevertheless recorded in the history of science as an event considered astounding by his own contemporaries.

According to Bartholin: "Iceland crystal is extracted from a mountain whose exterior is entirely made up of this transparent, water-clear substance, which tarnishes if it is moistened; this crystal calcines in a hot fire, and dissolves in concentrated acid." And, "objects looked at through it appear double when it is thick. [...] When the object appears double, one of the images is fixed, while the other moves in accordance with the movements of the diaphanous object."

The most famous Dutch scholar of the day, Christiaan Huygens (1629–1695), lost no time seizing on this observation. He turned it into an argument for his "oscillatory" theory of light, communicated to the Académie Royale des Sciences in Paris in 1678 and then presented in his *Traité de la Lumiére* [Treatise on light] published in Leyden in 1690. In it, he establishes a theory of propagation of the ray which was at that time called "extraordinary," namely the ray which does not obey the normal law of refraction. He presents a formulation of this theory as additional proof—following his explanation of reflection and simple refraction—that supports his idea of light as an undulating movement of the ether, proceeding by means of longitudinal oscillations in the direction of propagation.

Although for his part, the French mathematician and astronomer Gabriel de la Hire (1640–1718) derived from this a remarkable theory on the formation of gypsum in the vicinity of Paris, it would take more than a century for the science of crystals to regain possession of this discovery, wresting it away from both optics and mineralogy.

Another important discovery went unnoticed because it could not easily be integrated into the dominant interpretive paradigms, either cosmogonic or optical. The case is even more surprising because its author was one of the great names of science. In his monograph entitled *Réflexions sur plusieurs observations concernant la nature du gypse* [Reflections on several observations concerning the nature of gypsum] (1719), Antoine de Jussieu (1686–1758) established that this mineral consisted of parallelepipedic particles, all of the same shape. In doing so, he went against the traditional theory which stated that gypsum, like all crystals, was composed of thin "leaves." The mineralogist Guillaume-François Rouelle (1703–1770) soon made an observation that pointed in the same direction, discovering that particles of sea salt had a cubic shape.

The end of the century saw a proliferation of articles on descriptive mineralogy, which became a genuine fad. We are all familiar with this period of "salon science," with the popularity of "cabinets" of natural history and the spread of a mania for "collecting" within French high society. It has often been ridiculed, and the most picturesque pages of Bachelard's *La formation de l'esprit scientifique* [Formation of the scientific spirit] were drawn from several works of this period. But interest in what we now call the "Earth sciences" was definitely more than just a simple amusement: the nascent Industrial Revolution was now no longer on the ground floor, but had actually moved underground. Letters patent of June 11, 1778, established, "in a great room of the Mint at Paris, a chair of mineralogy and docimastic metallurgy" (docimasy being that branch of chemistry which determines the nature and concentration of useful metals in ores), in which "the professor nominated by the King shall give free and public lectures in this science." Thus was announced the creation of the first School of Mines in Paris (1783). But none of the work undertaken made the decisive contribution that would have answered the appeal embodied in the celebrated *Mineralogi eller Mineralriket* [Mineralogy or the mineral kingdom] published in 1747 by the Swede Johan Wallerius (1709–1785), and later translated into German, French, Russian, and eventually even Latin. The external shape of crystals was not always the subject of the kind of systematic study that, before Wallerius, had also called another very famous Swede away from his vocation: the naturalist Carolus Linnaeus (1707–1778), who devoted the third volume of his *Systema naturae* (1737) to a "taxonomy" of minerals.

At this point, history experienced one of those periods of sudden acceleration. Jean-Baptiste Romé de l'Isle (1736–1790), who had served as an officer in France's Indian campaigns in 1757, seized on Linnaeus' idea, took it seriously, and embarked on a classification of crystals based on their external form. In 1772, after years of toil, he published an *Essai de cristallographie* [Essay on crystallography] that marked a turning-point. He did more than just describe and classify crystals. He crossed swords with the ancient legends; taking up the torch already wielded by Bernard Palissy (ca. 1510–1589), he denounced the charlatanism inherent in pseudo-medical uses of crystals, and firmly rejected the theory that crystals arose from "coagulated" water, as well as the theory of "seeding." Above all, he discovered what would remain the first law of crystallography: the "invariability of dihedral angles," or the fact that in a particular chemical species, "the faces of a crystal may vary in their shape and in their relative dimensions, but the respective inclination of those faces is constant and unvarying."

In this, Romé de l'Isle had found an infallible classification principle. Although he allowed himself to draw from it some foolhardy conclusions about the chemical process of crystallization, he paved the way for a truly scientific crystallography which would emerge at first as formal and descriptive. It became a geometric science, backed up by routine use of the contact goniometer invented at his request by Arnauld Carangeot (1742–1806), a Parisian business administrator who studied under Romé de l'Isle and made significant contributions to the discovery of the invariance of crystal angles.

But it was the Abbé René Just Haüy (1743–1822), a pupil of Louis Daubenton (ca. 1716–1800) before assuming his mentor's position as professor of mineralogy, who may be considered the true father of the discipline. Regardless of what may have been said about him, Haüy started not from a simple observation, but from an idea: pondering the rhombohedral cleavage of Iceland spar, he was convinced that by analysis (by meticulously breaking up the crystal), he could identify the "primitive form" of the spar. He concluded from this that by repeatedly cleaving the mineral, one would end up, in every case, with a core configuration or "integrant molecule" around which the crystal was constructed. In 1784 he published an *Essai d'une théorie sur la structure des cristaux* [Essay concerning a theory of the structure of crystals] which stated two precepts: "There are two things for the naturalist to consider in crystals: firstly the shape of its constituent molecules; secondly the arrangement which they maintain among themselves." Haüy was soon made a member of the Académie Royale des Sciences, and continued to extend the depth and breadth of his research until his death. One might say that all modern crystallographers are his children. The disciples of Auguste Comte, for their part, believed that in his descriptive and formal approach they had found a stunning example of the progression of a positivist spirit—in opposition to earlier theological and metaphysical speculations.

But in retrospect, how can we help but emphasize the predictive power embodied in Haüy's ideas, as Françoise Balibar invites us to do with arguments borrowed from the very latest theories of physics? Haüy did not simply observe and formulate laws of geometrical regularity; he proposed a hypothesis concerning the microscopic constitution of crystals. If it could be observed, he wrote, the structure of a crystal would demonstrate that "what lies beyond the point at which I can no longer see resembles what I have seen up to now." Hence, "corpuscles" that would take the form of "rhomboids of extreme smallness." According to him, it is the geometrical arrangement of these corpuscles that is responsible for the specific position of the facets and edges and their angles. It is probably no exaggeration to say that with this enormously daring hypothesis, Haüy is now more than ever a presence in the "science of crystals."

Crystallography had thus conquered its subject and developed its methodology. It is often said that this was achieved by throwing off the tutelage of physics, chemistry, and the grand cosmological theories. The reader will now discover, amid a few detours, how the crystal as a scientific subject, escaping as it were from the discipline which had such difficulty catching it, has ended up turning contemporary science upside down, giving Haüy's hypotheses a meaning he certainly never imagined.

It would also not be out of place here to consider crystallography as an exemplar, in many respects, of the way in which science makes progress today: seldom automatic and always contested; not linear, but slowed by repeated and unexpected forks in the road. We will now see in action before our eyes, all around the crystal, the rational coordination of a number of research fields that once seemed irrevocably foreign to one another. We will also watch as technological progress (such as x-rays and synchrotron radiation) merges with theoretical advances to create new subjects for research, opening up new horizons for scientific investigation.

We will find that the "applications" of this science may overturn established economic and social preconceptions, if only a "decision" is made to use them. The question of how such a decision is to be made remains unanswered, since fundamentally, as we shall see, the question has never been asked. But Dr. Balibar's book ends with a more unsettling and even more serious question, one that demonstrates—if she will forgive me for saying so—a generosity of spirit unusual in this age of disenchanted egotism: the very fact that progress in crystal research has brought this field into the realm of Big Science now renders it practically inaccessible to countries which lack the necessary industrial resources. Surely it is here, rather than on the battlefield—or in the oilfields—that the destiny of our world is being played out. Three-quarters of humanity finds it has been stripped of the resources needed for any kind of creative research in the key areas of knowledge. These "developing" countries find themselves, at best, reduced to applying results obtained by others, in sectors imposed on them by the world market.

The spirit of research thus appears today to be a privilege of the well-to-do. Is it any wonder that its absence opens the way to that very fanaticism, born of despair, for which it represents the only cure?

Dominique LECOURT

I

PHYSICS MEETS THE CRYSTAL

THE SORROWS OF CRYSTALLOGRAPHY

For many years, physicists held the same view as the general public, believing that crystals were, to be sure, fascinating objects—still referred to by some as "magical," and quite striking when seen in museums—but doubting that the science of crystals could be anything but a small, distant province, of rather exotic appeal, in the vast continent of physics. When I began doing research twenty-five years ago, working in a crystallography laboratory was not very chic; the "in" crowd was over in solid-state physics, high-energy physics, and, more marginally, astrophysics. After all, wasn't crystal science still very closely tied to mineralogy, a fundamentally dusty and above all purely descriptive discipline?

No one saw any possible connection between these pursuits and physics, that science which for the last half-century had been revolutionizing human thought and life, and whose framework, if not its very soul, was the quantum theory. Everyone was wrong, as we shall see, because since at least the

1920s some of the leading theoretical physicists, pondering the structure of the laws of physics even as they made essential contributions to it, had already perceived the link and begun to study crystals. They were joined in this effort, moreover, by mathematicians specializing in group theory. But educational institutions, whose job it is to train large numbers of students, always lag a bit behind research at the cutting edge. And I must add in all fairness that the work I am referring to still concerned only a tiny fraction of the physics community.

We are taught in school that matter exists in three more or less ordered states: the gaseous state, in which molecules are free to wander wherever they wish within their allotted volume; the liquid state, in which loose bonds (and therefore a certain "order") are established between molecules; and the solid state, in which molecules and atoms (so we are told) are perfectly ordered and rigidly fixed in specified positions known as "sites." It is said that in this last state, molecules and atoms can just barely oscillate around their positions, (but not leave them) when the solid is heated, in other words when what is called the "thermal agitation energy" of the constituents is increased. This traditional description of the three states of matter has been called into question and considerably modified, since there exists a whole series of "intermediate" states (liquid crystals, glasses, etc.), investigation of which represents an advanced field of physics. The notion of equating a solid with a perfectly ordered microscopic structure has gone out of style; we therefore tend more and more to replace the expression "solid-state physics" with the more precise term "physics of condensed matter."

Since the 1920s, the celebrated quantum theory has been the theoretical physicist's instrument of choice. Established by the Danish physicist Niels Bohr (1885–1962) in 1913, it achieved its "complete" form in 1927, in the works of Werner Heisenberg (1901–1976) of Germany and the Austrian Erwin Schrödinger (1887–1961), among others, who raised questions of philosophical interpretation that still resonate both within physics and beyond its boundaries. The quantum theory actually arose in large part from concerns related to a science we now call solid-state physics but which did not yet bear that name. We can state without exaggeration that it was "invented" as a means of examining the disconcerting behavior of certain specifically "quantum" objects (that is, objects which differ from any reality that can be observed in the macroscopic world and do not obey the laws of classical mechanics), such as electrons. The most "mobile" of these quantum objects were found to be capable of influencing the physical properties of solid objects, particularly their electrical and optical properties. It is not generally known that Bohr, like Albert Einstein (1879–1955), spent the early part of his career addressing problems raised by the theory of electrons in metals. Bohr's thesis (defended in 1911) was in fact entitled *Studier over metallernes elektrontheori* [An investigation of the theory of electrons in metals], and we know from his correspondence with his first wife that Einstein, for his part, had spent a summer studying the first of the electron theories, which was developed by the German physicist Paul Drude (1863–1906) in the 1880s. Considering this, it is no wonder that the destiny of solid-state physics has been linked, right from the start, with that of the quantum theory.

Compared with this sparkling, dynamic research field, crystallography looked like a poor relation; many people dismissed it as a mere observational science. If this expression, as it is presently used, is understood to mean a science (like astronomy not too long ago) in which the scientist lacks the ability to modify phenomena on the basis of a previously constructed theory, it is true that mineralogy—in which "collecting," inventories, classifications, and catalogs constitute (or at least used to constitute) the essence of research activity—did fit the definition very well. And since crystals were the object of detailed descriptions along those same lines, crystallography, by contiguity and assimilation, appeared to be a particular, albeit intriguing, instance of mineralogy.

This point of view, widely held in the community of "noble" physicists, was in fact based on a misunderstanding of the history of crystallography. Those who subscribed to it had forgotten that crystallographers had been the first to look "with the eyes of the mind" into the interior of matter, where the eye of the body sees only an opaque surface. It is worth remembering at this point that as early as 1781, Father Haüy* had formulated the hypothesis that each type of external geometrical crystalline shape corresponded to what he called an "integrant molecule," on the basis of which the crystal was built up by successive layering. Haüy thus offered one of the very first proofs of the discontinuous nature of matter, at a time when such discontinuity was by no means unanimously accepted.

Nor is it true that because crystallography concentrated its attention on the configuration of external crystal surfaces, and based the classification of crystals on a methodical study of that configuration and its variants, its research never really focused on the intimate structure of matter. Because precisely what this science has been doing, since the time of Haüy, is to oscillate endlessly between the observ-

* We should not forget that crystallography is the child of this austere priest who proved sufficiently adroit to participate in the work of the Revolution while at the same time refusing to swear allegiance to it.

able macroscopic world (crystal faces and their external morphology) and the microscopic (the arrangement of atoms).

A typical example was the work of scholars such as Auguste Bravais (1811–1863) in France, or J. F. C. Hessel and M. L. Frankenheim in Germany, who established a rigorous classification of crystals based on various possible arrangements of atoms. In so doing, they demonstrated that the number of such arrangements is limited, and that they can be categorized into 32 classes (or "point groups") and 14 "figures." Each of these microscopic arrangements corresponds to a plane with a high density of atoms (hence the term "atomic planes"), which constitute the actual cleavage planes of the crystal, in other words the planes along which the material will readily split.

It is therefore incorrect to regard crystallography as a science of mere observation and classification, functioning solely on the basis of appearances and external criteria. For example, Gabriel Delafosse (1795–1878), a pupil of Haüy, could write in 1843: "This [atomic] structure, as we are considering it here, appears without question to be the primary characteristic in crystals, that which dominates all the others. External form, which until today has been privileged to absorb all the attention of crystallographers, is in our eyes now but a secondary characteristic, of importance only because modifications to it are always found to be subordinated to the particular laws of the internal structure and of the molecular constitution of the crystallized body. Seen in this light, crystalline form is indistinguishable from other physical properties...."

Nevertheless, crystallography as most commonly practiced until about 1960 represented the epitome of an inductive science, proceeding by moving back and forth between macroscopic observation and microscopic modeling. But although in the 1960s it was still being considered a "non-theoretical" science, that was essentially because it was regarded as "classical" (meaning, in the contemporary view, "non-quantum").

There is no denying that the field of quantum physics, even with its accompanying loss of *Anschaulichkeit* (graphicness) that was such an embarrassment to its creators at the time, has something about it that seduces great minds. The fact that it is impossible to represent electrons is, after all, pretty exciting; and the ability to overcome that handicap by "talking math" is nothing short of sublime. So crystallography, which sets as its objective the down-to-earth task of representing the position of atoms within solids, seems by comparison rather unambitious. What is more, when reduced to a study of the arrangement of the constituent atoms of a crystal, crystallography seems quite "inert" alongside the revolutionary science of solid-state physics, whose goal, despite all the familiar difficulties and triumphs, is to define how electrons make the transition from one quantum state to another. Confronted with a science that sets as its goal no less than the evolution and transformation of things in time and space, in an immense variety of circumstances, crystallography appears to be preoccupied only with nitpicking considerations of simple geometry.

Once again, it was wrong to set up such a clear-cut opposition, since no crystal is as "inert" as the rectilinear coldness of its faces might suggest. The vibrations of its atoms about their equilibrium positions are by no means negligible from a physical point of view; they seem so only in the context of an approach that professes to be purely geometrical. Moreover, it is through these vibrations, called "phonons," that the connection was made between solid-state physics and crystallography, with each science having something to say on the subject.

But theory is not everything; the value of a science can also be measured by the fertility of its applications. So how can one fail to notice, once again, the contrast between basic research in solid-state physics, the applications of which would revolutionize the very foundations of the world economy, and observations which seemed attractive only on an intellectually aesthetic basis and were not felt (or predicted) to be of any practical consequence? It is on solid-state physics, after all, that all of modern electronics and its innumerable applications are based: radio, television, cameras, pocket calculators, "surgically" accurate laser-guided bombs, etc. But what exactly is the science of crystals "good for"?

A final note to make matters even worse: William H. Bragg (1862–1942), one of the founders of contemporary crystal science, had the idea (original and shocking for its time) that women could possess intellectual qualities equal to those of men: he had therefore offered employment to women in his laboratory. Having thus defied custom, crystallography came close to being regarded as a "ladies' science," and there were endless jokes about English laboratories as feminized islands in an otherwise almost exclusively male universe.

Such, until quite recently, were the trials of crystallography.

X-RAYS: SHADOW OF A BREAKTHROUGH

But Haüy, Romé de l'Isle, Delafosse, and the other founders did have their posthumous revenge, because this inequitable situation is now a thing of the past. Without any sudden revolution or sensational discovery to catapult it to prominence, crystal science has achieved respectability. If I may say so, there is considerable irony in the fact that crystallographers now find themselves at the very core of physics research. Scientists who once disdained the discipline, and believed that they would never work on anything but solid-state physics, now realize that they cannot avoid doing at least a minimum of crystallography.

This is a wonderful topic for epistemological speculation, if one wants to investigate not only the formal structure of theories, but also the history of how problems are formulated: the apparently airtight partitions between these two methods of inquiry (exquisitely refined predictive theory on one side, and induction based on "pure" observation on the other) have been demolished. Crystals, it seems, have invaded science.

This conquest proceeded on several fronts simultaneously, and first of all in the context of theory. All "classical" crystallography was founded on the abstract idea of the "perfect crystal," to the point where crystals had to adapt to the requirements of theory; in other words, be made pure enough so it could be comprehended and discussed. But we shall see that crystals gradually established themselves as real objects, with all their defects and impurities, obliging theory to accommodate itself to reality. The second front was staked out in biology. Crystallography today represents one of the most frequent and active points of contact between physics and the life sciences, ever since the advent of what is called "molecular biology," a field to which it has made such decisive contributions. These inroads by crystallography into various branches of scientific activity are now extending as far as astrophysics. Simultaneously, crystallography has entered into an alliance with mathematics, the queen of all sciences: group theory has become its favorite theoretical tool, and crystallographers and mathematicians have discovered certain common interests. As to the economic importance of applications of crystal science, no one today would dream of underestimating them: there would be no miniaturization of electronics without crystals, and (as we shall see) no "advanced" materials without crystallography.

It has been, without question, a slow evolution rather than an abrupt revolution, but the groundwork for this crystallographic imperialism was laid a long time ago. Before invading physics, crystallography began by adopting its methods. It actually became a branch of physics on the day in 1912 when the German scientist Max von Laue (1879–1960) conceived the idea of studying the arrangement of atoms inside a crystal by using a type of radiation whose nature had remained a mystery for some time (hence the name "x-radiation") but which had been known since the experiments of the German physicist Wilhelm Roentgen (1845–1923) to be electromagnetic, just like light and radio waves. This radiation is characterized by extremely short wavelengths, on the order of 0.1 nm (or one angstrom unit, named for the Swedish physicist Anders Jonas Ångström [1814–1874]). What was the advantage of x-rays? As everyone now knows from their medical applications, they pass through most solids, and can therefore be used to "see" inside things. But why should they constitute the ideal tool for looking at atoms in solid matter? Very simply because of a general physical principle which states that, in order to see details of a certain dimension, they must be illuminated with a "light" that has a wavelength on the same order of magnitude as the size of the details being observed. It so happens that the distance separating the atoms in a crystal is in fact on the order of 0.1 nm; in a crystal of table salt, for example, the distance between the atoms (or, more accurately, between the Na^+ and Cl^- ions) is 0.48 nm. X-rays are therefore well suited to observing the respective positions of atoms inside a crystal.

I should point out that the "principle" I have just referred to is not, strictly speaking, a principle at all, since it is demonstrable. It is one of the fundamental results of the physical and mathematical theory of wave diffraction. As such, however, it is one of those theoretical statements that physicists, both theoreticians and experimentalists, are continually aware of—guidelines to which they refer almost unconsciously when analyzing a concrete situation. I would even go so far as to say that one cannot become a physicist without having internalized these few principles to the point that they become as natural as "two plus two makes four."

Let us look at a few similarly important principles by way of example. One is the fact that the hotter something gets, the faster it moves; a result of the kinetic theory of gases and more generally of statistical mechanics, which establishes a directly proportional relationship between the temperature of an object and the thermal agitation of its constituents. Another is the fact that constructing a phenomenon of short duration requires a wide spectrum of frequencies, a result of the transformation theory of Jean Baptiste Joseph Fourier (1768–1830). The half-abstract, half-concrete formulation of these statements betrays their intermediate status: they are the filters through which physicists abstractly interpret the concrete world.

But let us return to crystallography and to Laue. In 1912 he performed an experiment in which x-rays were diffracted by a crystal (or, more precisely, by the atoms in the crystal), thus enlisting physics—in this case electromagnetism and the wave diffraction theory—to assist in the observation of crystals (or, more specifically, of their internal structure). It is important to note that until then the process had gone much more the other way: crystals had been put to work in the service of physics, as a means of studying light. Just think of all those "crystalline optics" experiments devised during the 19th century.

What does a "Laue pattern" look like? It looks like a collection of spots, arranged on the photographic image in a way that obeys certain laws of periodicity. Someone who knew nothing about the process might think that it was a photograph of the atoms themselves. In reality, each spot corresponds to an assemblage of "atomic planes," in other words spatial planes containing a high density of atoms. If you imagine a periodic arrangement of points in space, you can easily "see" that it is possible to select certain planes, parallel to one another, which contain more atoms than others. A point or spot on a Laue pattern represents an orientation of planes. As far as the photographic record itself is concerned, it depicts all of the variously oriented atomic planes that constitute the crystal. It is therefore incorrect to interpret this pattern as a true photograph of the arrangement of the atoms. Nevertheless, the periodicity exhibited by the pattern is not without concrete significance; quite the opposite, it says something about the symmetries of the crystal itself, or more accurately about the arrangement of its constituent atoms in networks. So a Laue pattern actually does look like a photograph of the crystal, but a photograph in a space different from ours, in what crystallographers—and, following their lead, physicists—refer to as "reciprocal space."

For more than half a century after Laue did his research and presented his results, x-rays remained the tool of choice for studying the structure of crystals at the atomic level. They were used to "realize" the structures of most mineral compounds and certain organic compounds (the term "realizing a structure" is understood to mean determining the atoms which form the pattern whose repetition constitutes the crystal). These structures are listed in a sort of catalog called the *Structure Reports*, which give, for each known compound, the distances between atoms, angles between atomic bonds (the term used to describe the line segment connecting two atoms in a graphic representation of the network of atoms, or in those stick-and-ball teaching models exhibited in certain museums), the length of those bonds, etc.

For a long time, the principal activity of a crystallographer was to prepare and update this catalog. After a while, as still unknown structures became increasingly scarce, this activity lost its dynamism. It is only a slight exaggeration to say that there came a time when "publishing a structure" meant simply selecting a compound with a structure that was already known, changing a few of its constituent atoms by replacing them with another element in the same column of Mendeleev's periodic table, and making a Laue pattern of it.

THE AGE OF UPHEAVAL: NEUTRONS AND SYNCHROTRON RADIATION

Crystallography thus passed through a period of bureaucratization. However unkind the word might sound, it is not unjustified: an international administration did in fact arise, with its principal activity limited, for decades, to watching over the *International Tables*.

But everything changed with the advent of two new technologies applied to the study of the internal structure of matter. These have also helped, each in its own way, to create a link between solid-state physics and crystallography by shaking the latter science out of its doldrums. Bureaucratization would thus be followed by an era, if not exactly of "revolution" (I have already mentioned that things happen rather quietly in crystal science), then at least of upheaval.

The first upheaval came with the use of neutron diffraction as a means of studying crystal structures. This represented the introduction of quantum physics in crystallography. By directly applying one of the essential formulas of quantum theory—the statement by Louis de Broglie that every moving particle was associated with a wavelength (called, unsurprisingly, the "de Broglie wavelength")—someone got the idea of studying matter not just with photons (x-rays *are* photons, just like light, of which they are only a particular form), but also with particles of matter whose "wave-like" aspect, implied by de Broglie's discovery, would thus be exploited. These particles were diffracted by atoms, just as the electromagnetic waves associated with x-ray photons are diffracted. The basic idea, in short, was the following: since there is no fundamental difference, as far as quantum theory is concerned, between photons and particles of matter (such as neutrons), what is accomplished with photons (in this case diffraction by the atoms of a crystal) can just as easily be done with particles. All that is necessary, in a context where the underlying phenomenon is one of diffraction, is that the wavelengths associated with the two types of object be of the same order of magnitude. This, then, was the moment at which crystallographers got on the quantum bandwagon.

The neutron, discovered in 1932 by the English physicist James Chadwick (1891–1974), is a particle: it is one of the two basic constituents (along with the proton) of the atomic nucleus. The reason for the name "neutron" is that it has zero electric charge. But at the same time, for the reasons just discussed, it can be used as a wave. It is therefore quite natural to talk about the "wavelength" of a beam of neutrons. In France, the first neutrons produced for the specific needs of crystallography were created at Saclay during the 1960s. Then, as the scientific community's need for neutrons for crystallographic purposes assumed increasing importance, the joint French/German *Institut Laue-Langevin* (ILL to insiders) was built at Grenoble, designed to produce and use neutrons. Producing these neutrons actually involves almost no major difficulties at all, provided sufficient financial resources are available (hence the idea of a French/German collaboration). In this case—for once—Nature seems to have made life easy for the physicist: the requisite neutrons are produced at room temperature, and do not require very high or very low temperatures (which obviously would complicate the production process). That is hardly surprising, since we know that in general each temperature has a corresponding specific "agitation" (this being the famous principle that "the hotter something gets, the faster it moves"), in other words a specific energy and thus a specific velocity. Since the de Broglie formula states that the wavelength associated with a particle depends on its velocity, the wavelength of a neutron, in this case, consequently depends on the temperature at which it is emitted. It just so happens that the wavelength at which it is possible to "see" the atoms in a crystal—on the order of 10^{-10} meters—is associated with a velocity which in turn corresponds to about room temperature.

This innovation represented an initial leap forward for contemporary crystallography, since neutrons have the advantage of being... well, neutral (from an electrical point of view). They are therefore not influenced by the electrons surrounding the nuclei of the atoms in a crystal, and thus provide a "direct" view of the movements of the nuclei around their equilibrium positions, devoid of secondary effects that need to be subtracted. As a result, it is even possible to examine, much more readily than with x-rays, the phenomena I referred to a little earlier as "phonons." Neutrons are also more capable than x-rays when the atoms being inspected are "light" in terms of atomic weight: as soon as neutron diffraction became available, it was finally possible to do precise crystallographic studies of hydrogen atoms. Bearing in mind that hydrogen is one of the essential constituents of molecules of "biological interest," it is obvious that this stage of crystallography represented enormous progress.

One should not conclude, however, that neutron diffraction caused any upheaval in crystal theory: it did not become a quantum theory just because the tool being used to investigate crystals was a quantum tool. In fact, researchers continued to use theories developed on classical foundations early in this century; the basic phenomenon was still wave diffraction. Whether this effect is associated with photons or neutrons is of little consequence: all that counts is the order of magnitude of the wavelength compared with the scale of the atomic distances. Although a neutron is of course a quantum object—*the* quantum object *par excellence*, one might be tempted to say, since it clearly exhibits wave–particle duality—it participates in this case solely as a tool, and its quantum nature really has little to do with how the results are interpreted. In a way, one might say that although neutron diffraction brought crystallography close to the cutting edge in physics, its proximity was only superficial; although crystallographers have come within range, their theory has remained outside the quantum fold.

What is more, trying to construct a completely quantum theory to account for how x-rays or neutrons are diffracted by a crystal would be like killing flies with a sledgehammer. For example, it is illusory and thoroughly useless to attempt to build such a theory based on what is called "second quantization formalism," by studying the "fundamental" interaction between a photon (or neutron) and the ensemble of electrons and ions that constitutes the crystal. Let me take this opportunity to dispel the incorrect notion, so widely held among non-scientists and inexperienced physicists, that a quantum approach to a phenomenon will always make things more understandable than a classical theory. There are cases—and diffraction of neutrons and x-ray photons by a crystal is one—in which discussions of the principal phenomenon (in this case wave diffraction) involve only a single characteristic, namely, the "wave-like" nature (to use this somewhat quaint old term) of the object in question, and have nothing to do with quantum theory.

Quantum theory also has nothing to do with the second innovation mentioned earlier, namely the introduction of synchrotron radiation. This is a byproduct of what is called "high-energy" physics, that branch of the discipline which best embodies, in the eyes of the general public, the idea of contemporary Big Science, the kind symbolized by the great CERN accelerators in Geneva, or the proposed new Superconducting Supercollider in Texas.

The word "synchrotron" sounds intimidating, but the actual hardware is relatively simple, at least in principle. When physicists study high-energy particles, they do so by accelerating certain particles, especially electrons and positrons (the antiparticles of electrons), around a ring-shaped structure that can be as much as several miles in diameter. One of the fundamental laws of classical electromagnetism states that when a charged particle is accelerated, it radiates electromagnetic energy (or light, if you will). The more strongly the emitting particle is accelerated, the shorter the wavelength of the radiation and therefore the greater its energy. This is why x-rays corresponding to very short wavelengths can only be produced with highly accelerated particles, and thus by very powerful accelerators.

This makes it possible to obtain—or "harvest"—x-rays along the edges of an accelerator. It therefore became common practice for physicists interested in using x-rays for diffraction purposes to set up their equipment in little huts located next to an accelerator while the great and noble experiments proceeded. But this practice of "squatting" alongside Big Science became an annoyance to everyone, including the high-energy physicists and the crystallographers, and almost everyone in the scientific world therefore decided to build instruments specifically designed to produce "synchrotron" x-rays.

Crystallography thus gradually moved from the status of a barely tolerated squatter to that of a home-owner. Instead of sharing "beam time" with other users—as is still the case at DORIS in Hamburg (built in 1973) or at CHESS at Cornell University, and as things were at DCI in Orsay (France) until it was converted in 1985 into a machine set aside for crystallographers—the preference these days is for "dedicated" sources, such as the SRS source at Daresbury in Britain (inaugurated in 1981), and the Photon Factory at Tsukuba, Japan's new Science City (in operation since 1982). The same will be true for the European source scheduled for opening in 1993 and currently under construction at Grenoble.

The introduction of these new x-ray sources represented, for crystallography, an event comparable to the advent of the laser in the field of optics. Not only are they far more powerful than conventional sources—just as the laser led to an enormous increase in the power of light sources—but they also have certain characteristics completely unlike those of previous sources. One especially useful property is the ability to produce a "white" beam, meaning one that comprises an entire continuous range of wavelengths; by their very design, conventional sources were inherently unable to produce strong intensities at more than one or two wavelengths. Researchers can therefore now utilize sources with adjustable wavelengths, the applications of which are of course turning out to be much more varied than those of conventional sources. This capability, along with continual improvements in detectors, has made possible some important advances, including determination of the structure of proteins.

But once again, these technological upheavals represent changes only in the instruments of observation. Nevertheless, they have had very significant repercussions on the "social" status of crystallography: formerly a "minor" science, it is now asking for major pieces of equipment and large investments.

These days, and in our world, that is already the mark of a certain respectability.

This sort of upsurge within the stock market of the scientific community—which is not unconnected to the real stock market—would not be worth a book of its own, no matter how slim. At the same time, however, there has been a profound transformation in the ontological and theoretical status of the crystal, one which injects some real excitement into the future of crystallography, explains its present status, and offers a glimpse of its prospects.

THE NEW CRYSTAL

The theoretical foundations of the study of crystals remained amazingly stable from the early part of this century almost to the present day. A single theory—diffraction—was used to interpret observations (which were generally photographic images). I must emphasize here the fact that the process of diffraction, in which a wave that encounters an obstacle is scattered in all directions, can occur in a perceptible manner—assuming multiple diffracting sites, as is the case here (remember that the number of atoms in a crystal of ordinary size is on the order of 10^{23})—only if those sites are regularly arranged in space. Otherwise the result is merely a blur: a uniform, undifferentiated shading of the photographic plate.

This blurring effect is easy to understand by considering a phenomenon familiar to every physics student: interference. We know that superimposing one beam on another does not necessarily produce reinforcement of the light; it can also yield areas of shadow. It follows that if the number of beams involved is not two, but is equal to the number of atoms in a piece of crystal, a similar phenomenon will occur, except that it will be more localized. The observation screen (or the photographic plate), instead of containing equal numbers of dark and light areas, will exhibit very narrow and very intense areas of light, separated by large stretches of total blackness. The sensor will collect light only in certain very specific axes. This is exactly what happens when an experiment similar to Laue's is performed, and the spots seen on the image do in fact represent those axes of light striking the photographic plate. It is also evident that if defects are introduced into the periodicity of the crystal, the phenomenon will be blurred or fogged, just as interference will smear out the reception of a radio broadcast. Take the extreme case of glass, a material whose technological significance is continually growing. In glasses, unlike crystals, atoms are arranged in a disordered fashion (glass is in fact a liquid that has been abruptly cooled, and for that reason has retained the disordered structure characteristic of liquids). It is impossible to study the internal structure of glass by x-ray diffraction (or diffraction of any kind of wave): you will see absolutely nothing.

Crystallography has therefore been associated for a long time with the idea of the "perfect crystal," a crystal having no defects of any kind, no lack of periodicity, no fractures, no slippage between one atomic plane and the next, etc. Obviously, since perfection is not of this world, the perfect crystal is merely an ideal, and real crystals are all more or less imperfect. In fact, as long as the density of these imperfections remains within reasonable limits (glass being an unreasonably extreme example), the defects will not keep crystallographers from their favorite task: determining the structure of the crystal. The researcher can proceed as if the imperfections did not exist, and deduce from the results observed on their Laue patterns (for example), the three-dimensionally periodic structure that the crystal would have if it were perfect. Even better: in a nice irony of physics, this ideal crystal structure is easier to establish if the crystal is not *too* perfect!

For some time (although we shall see later that circumstances are now changing), crystallography has therefore found itself engaged in the curious process of describing, on the basis of experiments on real objects, what those real objects would be like if they were not real but ideal and perfect, and as if the idea of an impeccably ordered atomic arrangement had some reality.

Ever since Aristotle, the idea of perfection has evoked the idea of indestructibility, two characteristics which the philosopher attributed to a particular geometrical form, namely the sphere. From this arose, by generalization, a prejudice which stated that pure geometrical forms constituted a realization of perfection. We find this prejudice in Kepler. Galileo (1564–1642) rebelled against it under the banner of what he conceived of as continuity: his idea of the continuous turned him away from a belief that in Nature, where absolutely exact measurements are impossible, a continuous quantity such as the orientation of a surface could have an extremely precise value (as in the case of a geometrical polyhedron). One might in fact say that Galileo's physics was constructed specifically in opposition to the idea that pure geometrical figures could be realized in Nature. This runs counter to a certain traditional depiction of Galileo, but in any case it would explain why the science of crystals has developed alongside physics rather than as part of it. What is certain, and is confirmed by the texts, is that Galileo rejected crystalline perfection, and indeed found it repellent. One need only read the admirably poetic and delightfully ironic passage in which he refutes the idea that perfection should be associated more with immutable crystals than with mud and earth. It is only, he writes, because of their rarity that crystals are so highly esteemed; if earth were as rare, there would not be a single prince who would not spend millions to obtain a little of it, just enough to grow a jasmine or an orange tree in a pot. As for those who are silly enough to think that crystals are more perfect than soil or blood, they deserve to be changed into a diamond crystal by some Medusa, "to make them more perfect than they are."

These astonishing words bring us back to the change that crystal science has recently undergone, because what matters today to crystallographers is essentially the crystal just the way it is, with all its defects and imperfections. Not that perfection is not sought—quite the opposite: certain technological applications, especially the manufacture of electronic components, require more and more perfect crystals. But the point is that they are sought after as real crystals, not as some theoretical ideal. In general, for the past several years, it is only a slight exaggeration to say that the greatest appeal of a crystal (in the eyes of a crystallographer) lies in its defects. One might even go so far as to say that as this new interest has developed, crystallographers have become full-fledged physicists. That at least is the conclusion that comes from subscribing to Galileo's ideas, since his opposition to the idea of crystalline perfection originated, as he clearly stated, in his notion of physics. What explains the immutability of an object, he wrote, is not the fact that it has this or that shape prescribed by geometry, but rather the forces which act on its surface. The whole problem, he believed, was to understand which forces were causing a physical quantity (in this case the density of matter) to undergo a sudden discontinuity (from the interior of the crystal to its exterior, across the crystal's surface).

And the precise concern of crystallographers today is to achieve a true understanding of the forces involved in producing a crystal: both those which compel it to exhibit a globally perfect and ordered atomic arrangement, and those which create defects. From this point of view, there can be absolutely no doubt that—as Galileo had clearly understood—the first defect that must be comprehended is the surface of the crystal itself.

It might certainly seem odd to refer to the surface of a crystal as a defect. Still, in terms of the idea of a perfect crystal—namely a three-dimensionally periodic arrangement whose periodicity is uninterrupted, and which therefore extends to infinity in all directions—the very existence of faces which define the extent of the crystal constitutes a "defect." The reason why a perfect crystal must, almost by definition, be infinite in all directions, is what we might call a mathematical one, which in turn has to do with the existence of a powerful analytical tool known as harmonic development or the "Fourier series."

The word "harmonic" obviously has certain implications that evoke a musical concept associated with the idea of perfection. Similarly, the only way to analyze diffraction patterns, and hence to describe the ideal perfect crystal, is within the mathematical context of the analytical technique invented by Fourier. In this analysis, to which Bachelard devoted an entire book, the fundamental concept is that of periodicity; every function is resolved (or decomposed) into periodic functions of a particular type, namely the cosine and sine functions familiar to every schoolchild, which happen (as if by accident) to be called "harmonic functions." These functions, however, as anyone can demonstrate by imagining a sine wave, never begin and never end: their extension is infinite. If we acknowledge that harmonic functions are one way of representing ideal perfection, we see that the simple act of reducing the spatial extension of a magnitude (the density of matter, for example) to a finite domain forces us to "truncate" its mathematical representation; and that very act robs it of its claim to perfection.

The manner in which real crystals are formed has thus become an object of study for both crystallographers and physicists, and the two professions thereby lose their specific characteristics. The basic question is obviously that of "crystal growth"; the very metaphor of a crystal "growing" wipes out with one stroke the image of the crystal as an earthly embodiment of perfection, a cold image of death.

The meticulous study of crystal growth has turned up some completely unexpected facts. It has been found, in particular, that under certain very specific (and very unusual) conditions, it was possible to create transitory "round" crystals that had almost no faceting. Unlike the usual situation in which a crystal grows row upon row of atoms, forming terraces whose alignment is defined by the order which prevails within the body of the crystal itself, in this case that order is no longer felt, and the surface develops freely, released from the constraints imposed by the substrate. This strange phenomenon should eventually shed some light on the more general problem of the energy of a crystal's internal cohesive forces, and on the way in which they are released at the surface. The same type of information should be forthcoming from relatively recent studies of "surface reconstructions" observed on silicon surfaces by tunneling electron microscopy, a technique invented in the mid-1980s that allows us to see, truly and individually, the way in which atoms are arranged at the surface of a crystal.

Consider how far we have come since the time, not so long ago, when it was possible to take sides for or against the "atomic hypothesis," when the opponents of that hypothesis claimed that atoms were mere fictions since it was and would always be impossible to see them. These days it is difficult to believe that the work of Jean Baptiste Perrin (1870–1942) and of Einstein on Brownian movement was being trumpeted, until almost 1920, as proof of the existence of atoms—an existence that today is such an integral part of our collective imagination.

II

CRYSTALS IN PHYSICS

THE VALUE OF DEFECTS: SUPERCONDUCTORS AND "DESIGNER DRUGS"

The reason that crystallographers today are no longer afraid to look crystals "in the eye"—accepting them as they are with all their defects—and that they are actually eager to study those defects is because they have managed to develop observational techniques (using x-rays or neutrons) that go beyond a strict application of Laue's "historic" method.

The earliest of these new techniques was called "x-ray topography." It involves examining on a finer scale what goes on in a Laue spot, and thus isolating one spot among many similar ones. It can be demonstrated that under these conditions, the intensity of the x-rays (or the neutron wave) that passes through the crystal is distributed within a triangle known as the "Borrman triangle." Let us now postulate that a crystal contains some defect, say a dislocation line—a fault in the triply periodic arrangement due to the fact that a semi-plane of atoms is suddenly missing (this type of defect occurs quite often as a crystal grows). If the dislocation line crosses the Borrman triangle, it is in a certain respect "illuminated" by the x-rays; it is then obvious that a phenomenon similar to the one which causes a shadow will then occur, and that the "shadow" of the dislocation can then be observed on the exit face of the crystal simply by placing a photographic plate very close to the exit face and parallel to it. Developed in France and in England in the 1950s and 1960s, this technique is now widely used to study a variety of defects; its capabilities have, of course, improved tenfold with the advent of synchrotron sources. The electronic components industry is a heavy consumer of "topographies," since the electrical properties of a material depend strongly on the type of defects it contains.

Another method for studying defects, dating from approximately the same period, is electron microscopy, which again offers a way of mapping the imperfections in a crystal. In addition to biologists, whose extremely thin samples are well adapted to the constraints of electron microscopy, it is used primarily by geologists and mineralogists.

A newer technology is one called EXAFS (extended x-ray absorption fine structure), which can be used to determine the precise environment of a foreign atom located in the crystal. This highly sophisticated technique can only be used if a synchrotron source is available. Its application is not confined to crystalline matter (which is obviously a very important point), but extends also to the study of glasses, whose atoms are not arranged in a crystalline lattice and therefore cannot be studied by conventional crystallographic methods. Once again, geologists and mineralogists make assiduous use of these topographies. At the same time, moreover, mineralogy itself is experiencing considerable upheaval: it is not unusual today to find mineralogists, crystallographers, and solid-state physicists all patronizing the same accelerator.

Another advancement in experimental techniques carries the seeds of a true revolution in materials: instead of seeking out the natural crystals that are required, scientists are now constructing them at will, *à la carte*, as it were, atomic layer by atomic layer, and almost atom by atom. The technique used is called MBE, or molecular beam epitaxy. "Epitaxy" is a high-tech word combining two Greek roots, ταξι-, describing rows of soldiers drawn up in an army's order of battle; and επι-, meaning "upon"; and the technique of epitaxy in fact consists of "building" a crystal by arranging atoms layer upon layer, like soldiers on parade. This is done by propelling atoms, using a beam of molecules, against a suitably prepared surface, onto which they stick. The flow of atoms directed in this manner can now be controlled with sufficient delicacy to deposit them one layer at a time.

This kind of technological innovation cannot fail to excite the imagination. Just think back to the production of materials in the 19th century: metallurgy, the prototype of all materials fabrication processes, proceeded essentially by trial and error. A little more carbon, a little less iron, this ore rather than that one...and then wait and see what the practical results were. Now, on the other hand, it is possible to specify very precisely the properties that you want a material to have, and to determine the exact kind of atomic arrangement that corresponds to them. Then you simply make it. This sort of technological virtuosity obviously would not have been possible without advancements in crystallography, which specifically studies the arrangements of atoms.

The same example also illustrates the degree to which crystallography is now intimately linked to developments in solid-state quantum physics. After all, the reason for constructing materials in this fashion, atomic layer upon atomic layer, is not simply for the pleasure of performing technological stunts, or even for the benefits that can be derived from the technology adopted. There is also a theoretical reason: these materials can be used to study "quantum wells."

A quantum well is a theoretical object that will be familiar to any physics major. It is, in fact, a "potential well," which in turn means any region of space into which a particle (let us call it an electron) is strongly attracted. For the electron in this example, a proton—which attracts the electron because they have opposite charges—constitutes a potential well. In classical physics, when a particle "falls" into a potential well, it does not come back out: it would need to be supplied with additional energy to make it emerge from the hole. Such is not the case in quantum theory, in which the particle, even if it does not have the energy needed to get out of the well, can still partially escape from it. The reason is that quantum theory states that its probability density outside the well is not zero: there is a sort of probability leakage out of the well. This is one of the most astonishing effects of quantum physics.

Without getting into details, let us simply say that depending on the values of the parameters characterizing the quantum well, the circulation of electrons (which are the particles usually involved) has certain features. Conversely, if we want the electrons to behave in a specific manner, we simply need to calculate the parameters of the well to meet that need. It so happens that a stack of different materials of adjustable thicknesses—which is what is produced with molecular beam epitaxy—looks, to an electron, just like a series of "potential wells." Obviously it is possible under such conditions to produce any given material "on demand."

There are now no limits to this type of procedure, and crystal engineers have become veritable architects in miniature, especially in the pharmaceutical industry. It is possible today—or at least it will be tomorrow—to construct molecules practically atom by atom, in order to impart to them some particular therapeutic property. The slightly flippant term for such constructions is "designer drugs." For example, molecules that are assigned the mission of fitting into strategic locations in a molecule of the target protein can inhibit the pathological function of that protein. Current research is being focused on the construction of a molecule to inhibit the enzymes responsible for sleeping sickness.

This kind of work, of course, requires a great deal of computing power in order to visualize accurately the configuration of the target enzymes. Using a computer, it is possible to reconstitute the structure of an enzyme (or, in other cases, of a crystallized virus), locate atoms, and even identify the position of a certain type of atom. The computer is obviously also the means by which, using the same methods, researchers can find the "holes" that occur in the atomic configurations of enzymes. The display screen shows the three-dimensional structure of the molecule along with its recesses; it is then relatively easy to analyze the reactions that occur between one molecule and another, since these reactions essentially involve molecules "docking" into these recesses. If, therefore, the goal is to destroy a virus before it penetrates the host cell, becomes active, and reproduces, it is absolutely imperative to know the position of these "holes." Only with crystallographic methods is it possible to produce such maps. In general, determination of the crystal structure of organic and biological substances has become a fundamental prerequisite for studying them, just as with minerals.

For a non-specialist like me, it is odd to realize the importance of three-dimensional spatial factors in explaining mechanisms as subtle as those of the immune system and the body's defense against viruses. One might almost be tempted to consider that too simple, and it seems almost abnormal that the placement of atoms in space should be so important.

The fact that crystallography is so centrally located at the heart of fundamental and applied research in the life sciences should actually come as no surprise, since molecular biology itself is descended from crystallography. François Jacob quite rightly emphasized this in his famous book *La Logique du vivant* [The logic of life]: "x-ray diffraction had been used in Great Britain at the beginning of this century to analyze the organization of simple crystals, such as sodium chloride. From this arose a school of crystallographers interested in resolving the internal structure of a wide variety of compounds, including biological macromolecules, since these physicists were firmly convinced that the functions of the living cell had to be based on the configuration of these molecules. It was in fact one of them who proposed the term 'molecular biology' to describe this sort of analysis."

The process, as we know, proved to be a long and laborious one: it could not even begin without some real cooperation between crystallography and theoretical chemistry, and without a link between these research fields and a theory of heredity that was still merely formal. Back then, the biological revolution was still in the future; now it is essentially complete. It is amazing to realize that in the final analysis, thanks to widespread acceptance of the concept of the crystal in physics, that old analogy between crystal structure and the structure of living matter that was so dear to Buffon in the 18th century has now again become an operative principle, although in a way that Buffon would never have imagined.

One last word before we leave these new adventures in crystal science: it has to do with the implications of crystallography for the development and utilization of the new "high temperature superconductors." There can be no doubt that the technological revolution of the future—even though the word "revolution" is somewhat debased—will, if it ever occurs, be based on the manufacture of materials which are superconducting (that is, offer zero resistance to the passage of electric current) at "reasonable" temperatures, namely temperatures on the order of a few degrees Celsius (or about 290 degrees Kelvin on the absolute temperature scale). Development of these new superconductors would have considerable economic and even political consequences, since in our world, energy still represents a political factor of capital importance. According to some estimates, these new materials would lead to a reduction in global energy transport losses of almost 30%.

At present this is only a distant prospect, since these new superconductors become superconducting only when cooled to a temperature of about 90 K (a frigid −180°C). But hopes remain very high, since the discovery of these new substances is both very recent (1986) and the result of chance—one of those inevitable accidents that punctuate the history of science and technology. It is not inconceivable that, with suitable modifications to their structure, they might be usable at higher temperatures. All of these substances belong to related families: they are ceramics consisting of oxides of copper (Cu) and other elements, especially barium (Ba) and yttrium (Y).

Even since it appeared center stage, YBaCuO (pronounced "ibakuo," as if it were a Japanese prefecture) has become all the rage at physics conferences—and that includes crystallography meetings. Speaker after speaker reels off reports of structural studies (crystallographic structural studies, that is) of substances that look very much alike, one more boring than the last. In fact, in the absence of any theory of high-temperature superconductivity that might indicate where to look, physicists have for the moment been reduced to cataloging these structures in an attempt to come up with substances that might exhibit the properties of high temperature superconductivity.

Given their predicted economic impact, it is easy to understand that this kind of research is being vigorously encouraged, and that conferences and publications are top-heavy with its results. Industrial laboratories were first into the breach, with IBM in the lead; but since then everyone has joined in, and public funding is reaching very high levels. The paradoxical effects that usually result in such circumstances were quick to show themselves: everyone is keenly aware that in order to get money nowadays, you have to be working on high-temperature superconductors; so everyone has one or more projects or contracts "in the pipeline." Somewhat the same thing is happening in biology, where AIDS research is siphoning off considerable funding (although with greater justification).

The results of this mobilization are turning out, I must say, to be disappointing for the time being. Endless new oxides with superconducting characteristics are being discovered, but no theory has been advanced to explain them.

SYMMETRIES AND DISSYMMETRIES

One could extend *ad infinitum* a description of the instances where crystals as such (in other words, with all of their potential imperfections and complications) have integrated themselves into the fabric of the latest physical (and, as we have seen, biological) research. I would prefer, however, to re-examine some conceptual aspects of the crystal. From this point of view there is no doubt that the word (and concept) which immediately leaps to mind is "symmetry."

Right from the start, it is easy to understand that the idea of symmetry is linked to that of periodicity, and it is by way of the concept of invariance that a connection is made between these two notions. Here is a classic example of what physicists (and mathematicians) understand by "invariance": Assume a cube in which every face is identical. This cube is shown to a person who then leaves the room. The cube is turned so that it presents one of its other faces to the person as he or she returns to the room. That person will then be unable to tell whether anyone has or has not touched the cube. This demonstrates that the shape of the cube is "invariant" with respect to certain operations.

The same applies to periodicity. Let us take the more abstract but still familiar example of what is called a "sine wave," the periodic curve *par excel-*

lence as we saw in the context of the Fourier series. It is obvious that if it is shifted exactly one period over, an identical curve results. If no reference point is available, it is impossible to tell whether the operation performed (displacement, in this particular case) has changed it or not; in technical terms, we say that it is invariant with respect to displacement. This last example has the advantage of demonstrating what today constitutes one of the fundamental laws of physical theory ("fundamental" because it goes to the bottom, or foundation, of things): associated with every invariance (or with every symmetry, since the two concepts are linked) is the conservation of a physical quantity. Let us imagine, for example, that our sine wave represents the change over time of a phenomenon: the physical quantity associated with this particular invariance is obviously the periodicity of the change (the "period"), which always remains the same. All we need remember, then, is that in quantum mechanics, energy and frequency (which is defined as the inverse of the period) are associated by means of Planck's constant—the famous equation $E = h\nu$—to generate the inference that when a phenomenon is invariant with respect to displacement over time, energy is conserved, meaning that it always retains the same value.

This line of reasoning leads to a result, called Noether's theorem, which not only affects quantum physics, but has universal value: invariance in displacement over time is associated with conservation of energy. The movement of a planet around the Sun, which at a first approximation is invariant with respect to time (that being the meaning of the word "revolution" in this instance), is a phenomenon in which energy does not change its magnitude. What applies to time also applies to space: any phenomenon that is invariant with respect to space corresponds to a conserved quantity, in this case the quantity of movement (also called "momentum"). To be completely rigorous, I should add that all of this applies only to so-called "isolated" systems. Similarly, invariance with respect to rotation is associated with conservation of angular momentum (or kinetic momentum), etc. Even more fundamentally, these "conservation laws," which can thus be identified with invariances or symmetries, must be correlated with the assumed homogeneity and isotropy of space and time (or "space-time").

This is easy to understand in the case of space. Saying that a phenomenon is invariant in displacement in space means that it happens identically here or in Tokyo or on the Moon; in other words, that space is "homogeneous." This is obviously begging the question, no matter how sensible it might sound; but without these assumptions, it is quite simply impossible for a physicist to think or talk.

If I have gone on at some length about these general considerations, which relate to all of physics (both classical and quantum) and not just to crystallography, it is for the sake of the essential idea that I wish to emphasize in this book. Although it has been institutionally separated for some time from what we conventionally call "physics," crystallography has in fact always, since its origins, put into practice the most fundamental methods utilized by physics (taking "fundamental" in the sense of the word defined earlier). The idea of symmetry was already the daily bread of crystallography pioneers at a time when it had not yet achieved the place in physicists' minds that it occupies today, thanks essentially to developments in quantum theory. Just recall, for example, that current research on "elementary" particles is being based on considerations of symmetry.

In dwelling on symmetry, I am merely bringing up to date an idea first advanced in 1894 by Pierre Curie (1859–1906), or rather observing the extent to which that idea was prophetic. Expressed in the words of Curie himself, who was a crystallographer before turning to his better-known research on radioactivity, "I believe it would be valuable to introduce into the study of physical phenomena those concepts of symmetry so familiar to crystallographers." And he added: "Physicists often use the considerations raised by symmetry, but generally fail to define the symmetry in a phenomenon, since very often the symmetry conditions are simple and obvious almost *a priori*."

The fact that physicists at the end of the 19th century found themselves in the position of Molière's *bourgeois gentilhomme* with respect to physical theory—doing crystallography without realizing it—appears very clearly in the example of Newton's law of motion, the crown jewel of classical physics. This law establishes a relationship among three physical quantities: the force F exerted on an object; the inertial mass m of that object; and the acceleration a imparted to the object by the force. Remember that acceleration is defined as the derivative with respect to time of the velocity v, which varies (or can vary) over time. For Newton, this law represented a sort of axiom, and upon that axiom was constructed all of classical mechanics. But in retrospect—for us 20th-century physicists almost ninety years after the theory of special relativity (1905) developed by Einstein—this "law" looks much more like the consequence of an invariance principle. That principle, of a kind slightly more complicated than the ones mentioned earlier, states that the same laws of Nature hold under a special transformation, which involves putting the system on a train passing by at a constant velocity, so that the velocity of the train (V) is added to the velocity of the system (v). Although stated in this undoubtedly somewhat odd manner, this is the essence of the famous principle of special relativity, which the law of motion (like every law of Nature) must obey. And so it does, although Newton never consciously considered it. Specifically, neither the mass m nor the force F are modified in the operation in question; all that changes is the velocity of the system, which starts out as $v(t)$ and becomes $v(t) + V$, with the "(t)" indicating that v is a function of time. We know that in Newton's law, it is not $v(t)$, but its derivative with respect to time, that matters; but $v(t)$ and $v(t) + V$ (where V is a constant) have the same derivative, as expected. We can therefore state, using Pierre Curie's terminology, that if Newton had paid more attention to symmetries, he could have deduced his "law" from the super-law represented by the relativity principle as the principle of symmetry in Nature, rather than positing an axiom.

But let us not forget that Pierre Curie could not say that, since he was writing in 1894, and the theory of special relativity that revealed to physicists the heuristic importance of invariance dates from 1905. So it was impossible for Pierre Curie to walk through the landscape of the fundamental theoretical principles discussed here: he stated his ideas in the form of reflections on "cause and effect" relationships, and presented his research in the context of a general consideration of the principle of causality. It was, in fact, as an "efficient cause" that symmetry (or rather dissymmetry, as we shall see) appears in his research, and we might say that the conclusions he draws from it take the form of a modest "object lesson" addressed to physicists by a crystallographer.

Here, once again in his own words, are the causality "principles" as he formulated them: "When certain causes produce certain effects, the symmetry elements of the causes must be present in the effects that are produced. When certain effects exhibit a certain dissymmetry, that dissymmetry must be present in the causes that engendered it. The converse of these two propositions is not true, at least not in practice; meaning that the effects produced can be more symmetrical than the causes."

In other words, in very general terms, it is the medium that structures the effect. Things are more interesting at a more detailed level, and all of the subtlety resides in the definition of the word "dissymmetry," of which Curie also said, in a famous phrase, that it "creates the phenomenon." We must not forget that the symmetries in question are geometrical, defined as a combination of certain "symmetry elements," in other words certain geometrical operations (such as inversion with respect to a "symmetry center," mirror reflection in a plane, rotation through a certain angle about a given axis, etc.) which conserve the object. By "dissymmetry," Pierre Curie therefore means the absence of certain symmetry elements (with the understanding that it is possible to list all of the symmetry elements of a given figure).

An example will make it easier to understand the effects of this dissymmetry, which must not be confused with "asymmetry" (defined as the absence of any symmetry), and still less with "antisymmetry," which denotes a symmetry effect with a change in sign (and which, moreover, is not a geometrical operation). Let us consider a tourmaline crystal whose symmetry elements are familiar to crystallographers: a "third order" axis and three longitudinal planes of symmetry intersecting along this axis. The notation for this structure is (3 m). This crystal is the "medium," the "cause." If we heat it, it becomes polarized, in other words an electric field appears within it (the pyroelectric effect). This field is the "effect." Now, an electric field is characterized by certain symmetry elements: an axis of order infinity, and an infinite number of symmetry planes passing through that axis, written (∞ m). The symbolic notation makes it quite evident that the (∞ m) group contains the (3 m) subgroup. The appearance of the electric field with (∞ m) symmetry in a medium with (3 m) symmetry conforms to Pierre Curie's dictum: "The symmetry elements of the causes [in this case the crystal] must be present in the effects produced [in this case the field]." We would also not be surprised, if we had read Pierre Curie, to find that other crystals besides tourmaline are pyroelectric, and that all of them lack the same symmetry elements, meaning therefore that all of them have the same dissymmetry—which, moreover, is the same dissymmetry as that of the electric field.

This "negative" characterization of dissymmetry, defined as an absence, explains how theory deals with the impurities and defects of a crystal, or more simply how it manages to break out of the narrow confines of crystalline perfection. A defect induces what is picturesquely called "broken symmetry": a new absence of one or more symmetry elements. This in turn produces a phenomenon, following the principle that "the dissymmetry creates the phenomenon." As an example, let us look at the case of a "vacancy," the term for the absence of an atom at a site where an atom ought to be. In principle, the existence of the defect breaks all the symmetries of the crystal; if the defect is exceptional, however, meaning that there are not too many defects of the same kind in the crystal in question, the effect is imperceptible on a macroscopic level. But what if the vacancy (albeit a rare one) is located at a center of symmetry of the crystal lattice? All the symmetries of the lattice would then be lost, except the one corresponding to that center of symmetry; a dissymmetry (absence of all the symmetries minus one) would then appear, and that indeed creates the phenomenon, thereby making the defect observable.

This presentation is obviously crude and schematic, because in practice, theoretical lines of reasoning cannot be fertile and precise unless they are formulated in the context of group theory.

CRYSTALLOGRAPHY DISCOVERS GROUP THEORY

Indeed, the preferred theoretical tool of the crystallographer is quite obviously *group theory.* I say "quite obviously" because I imagine that most of my readers, thanks to the initiatives of primary education reformers, have imbibed group theory with their mother's milk, and are therefore not unaware that the *group structure makes it possible to transform a set of objects (elements) into that same set*—which is precisely what symmetry operations are all about. Those who were fortunate enough to begin their schooling after these reforms were instituted should therefore know that a group consists of a set of elements having a "product" (or "internal composition law," "internal" in this case meaning that the product of two elements is still part of the set). This product has a neutral element, which when multiplied by any element of the group yields that same element; each element of the group also corresponds to an inverse element (defined as the element of the group whose product when multiplied by the element in question is the neutral element). It is then clear that, when considered as a whole, the symmetry operations of a crystal constitute a group: the product of two symmetry operations is still a symmetry operation; the neutral element is the "identity," or the operation which consists in doing nothing; and there is an inverse operation corresponding to every symmetry operation.

Group theory can now be used to systematize and rationalize the symmetry concepts that were developed, without this tool, by crystallographers of the two previous centuries (Auguste Bravais, Louis Pasteur, and Charles Mauguin in particular). It also provides a better understanding of how a triply periodic atomic arrangement produces faces on the macroscopic level—in other words, how the external shape of something, called in German its *Gestalt*, is determined by the arrangement of its internal constituents. At the most basic level, group theory is in fact what has allowed a rigorous demonstration of Haüy's intuition concerning "integrating molecules"—which I will now try to explain.

To do so, we must distinguish between two types of symmetry in a crystal. The first type is macroscopic orientation symmetry, that of the faces of the crystal. This is called *orientation symmetry* (since it is the orientation among the faces which characterizes the shape of a crystal). In addition to this symmetry and distinct from it, we have what is called *positional symmetry*, or microscopic symmetry, relating to the spatial arrangement of atoms.

This latter symmetry encompasses all the operations that bring the atomic arrangement into coincidence with itself. These operations are of two kinds: displacements (shifting from one atom to another) and changes in orientation (rotations about a center or an axis); or any combination of displacements and rotations.

The correlation between positional (atomic) symmetry and orientation symmetry (applying to the faces of the crystal), corresponding to the transition from microscopic to macroscopic, can be made by retaining only those positional symmetry operations which correspond to changes in orientation, since the displacements corresponding to these atomic operations are imperceptible on the macroscopic scale. Demonstrating that the orientation symmetry reflects the positional symmetry therefore means demonstrating that one can, *by reduction*, move from the group of positional symmetry operations, called the "position group" for short, to the "orientation group." The reasoning proceeds in four stages:

1. Count up all of the elements in the position group.

2. Adapt the result to the lattice structure of the crystal; here one finds a demonstration of the famous theorem that a plane cannot be tiled with just any old polygon, and especially not with regular pentagons. This realization completely revolutionizes the way we look at wallpaper patterns, mosaics,

and fine inlay work; and also prevents those of us whose hobby is lacemaking from embarking on geometrically impossible patterns.

3. Bearing in mind this lattice constraint, count up all of the permitted symmetry operation groups. Note that the number of such symmetry combinations is finite: there are only 230 "space" groups.

4. Reduce these 230 groups to the orientation (or "point") symmetry groups, leaving out all the displacement components of the operations counted during the previous step. We then find that there are only 32 possible groups for the symmetry of a crystal's faces.

All that remains is to observe that every crystal in Nature does, in fact, fit into one of these groups—no surprise to anyone, given the "unreasonable effectiveness of mathematics in physics," in the words of Eugene Wigner, the man who has given the most thought to the introduction of symmetry concepts in quantum physics. Einstein, for his part, would probably have said: "What a pity it would have been, if God hadn't made it that way."

It is not enough to say that group theory enjoys considerable prestige in quantum physics; it is quite simply the essence of quantum physics. Once again, how can one fail to be surprised that it was not until the 1920s and 1930s that symmetry considerations, the subject of incessant study by crystallographers—

who, after all, were also dealing with Nature just like the physicists—found a use in physics? Hermann Weyl (1885–1955), the mathematician who began introducing group theory into quantum physics in 1928, just a year after the official birth date of quantum mechanics, simply notes, in the introduction to his famous *Gruppentheorie und Quantenmechanik* [Group theory and quantum mechanics], that for 60 years, group theory, regarded by Felix Klein as the most significant theory in the development of 19th-century mathematics, had no application in physics other than crystallography.

It is not my intention to explain how group theory produces results in quantum physics, but one can get a small and vague idea by considering the role played in quantum physics by the inherent states of certain physical quantities (energy in particular), in other words those states which exhibit a certain type of invariance. That will indicate why the famous "quantum numbers" cannot, in the final analysis, be anything other than the characteristics which represent a certain number of invariance groups. More modestly, let us see why (or rather how) group theory operates simultaneously in both quantum mechanics and crystallography. I suggest that this double role can be explained with reference to the problem of the discrete and the continuous. Claiming the "right to dream" even in scientific matters, I will take the risk of speaking freely in a context where, as we know, speech is always placed under the hegemony of a community of peers. I believe that such freedom has often proved fertile in the past.

In the quantum realm, group theory proves to be of inestimable help, in fact absolutely essential, whenever the *evolution* of a system under the action of an external agent is being studied. One of the great problems of spectroscopy, for example, is that of selection rules. The question can be stated as follows: Knowing the characteristics of the inherent states of a system (a certain molecule, for example), how can we explain why certain transitions from one of these inherent states to another are not found or observed in Nature, or, in the jargon, are "forbidden?" These forbidden states cannot be understood without studying the possible symmetries of the system as a whole (molecule + external agent responsible for the transition). Here is where group theory comes in. We find that an electric field (which, as we have just seen, has a certain symmetry) and a magnetic field (which has a different one), do not generate the same transitions: a certain transition that is forbidden when it is induced by an electric field is permitted when it is excited with a magnetic field. The molecule, a rigidly discrete structure whose energy levels have a symmetry corresponding to their respective quantum numbers, thus evolves in a completely different manner depending on the agent that is driving it. The evolution of the molecule is thus governed by the inherent symmetries of the molecule, since the *potential* transitions are entirely determined by these symmetries and by those of the external agent, since actualization of these transitions into *possible* transitions depends on the external circumstances, or in a way on the environment.

The same type of phenomenon, combining the potential and the forbidden, occurs in crystallography, especially when a crystal evolves or grows. As we have seen, the external shape of a macroscopic crystal, or more precisely its orientation symmetry or orientation group, is determined by its internal symmetries, one of the 230 position groups. We then find a *discrete* structure as the basis for the phenomenon. But that does not mean that *continuous* variations can be completely excluded from a description of the crystal. By placing the emphasis, as has been done so far, on the structure and symmetries of the crystal, we have temporarily neglected the "metric" of space, in other words the manner of determining the distance separating two points in that space—which can undergo any desired affine transformation (compression, expansion in one direction, etc.) with the sole proviso that it be compatible with the crystal's symmetry group. By means of the "metric," the continuous thus takes its revenge on the discrete, at the very heart of the realm of the discrete *par excel-*

lence: the crystal. This revenge is especially conspicuous in the phenomenon of growth, where the characteristics of the external environment are precisely what is at issue. It may happen that one face is prevented from growing: this circumstance does not change the (discrete) symmetry of the crystal, but acts on the "metric" of its elementary lattice (or cell). But the final result, as in the case of a molecule subjected to a field, is the actualization of a certain number of discrete *potentialities* (those which are permitted by the orientation symmetry and, more profoundly, the positional symmetry of the crystal). Here again, the specific potentialities that become actuality are determined by the symmetry of the external agent.

However "free" they may be, these thoughts still refer to a "great authority," more specifically to a paragraph in Hermann Weyl's book, *Philosophy of Mathematics and Natural Sciences*. He writes: "The visible characters of physical objects usually are the resultants of constitution and environment. Whether water, whose molecule has a definite chemical constitution, is solid, liquid, or vaporous depends on temperature. The examples of crystallography, chemistry, and genetics cause one to suspect that this duality is in some way bound up with the distinction between discrete and continuous. Here is one tentative suggestion. For a character like the (positional) symmetry group $\mathfrak{G}$ of a crystal that by its very nature (in agreement with an adopted theory) is capable of discrete values only, a specific one among these values is constitutive, whereas for a character with a continuous range, such as the character 'metric compatible with the given group $\mathfrak{G}$,' merely the range (here the pencil of all positive quadratic forms invariant under $\mathfrak{g} = \{\mathfrak{G}\}$) is constitutive. An individual stationary quantum state with its energy level is a good constitutional element (in spite of quantum jumps due to interaction); not so a wave state or, more generally, a statistical ensemble of wave states. Temperature is the environmental factor *kat' exochen*. And Weyl concludes, with some justification: "I think that this whole problem is in need of epistemological clarification."

In the course of his book, Weyl provides additional insights into what he is thinking. Speaking of the two types of constitutive factors—which for a crystal are its group membership and its "metric," one discrete and the other continuous—he correlates these with the space-time of general relativity. For him, the *nature* of the metric (which we know to be locally Euclidean in general relativity because of the "equivalence principle") can be regarded as identical at any point (regardless of the distribution of masses). The coordinate system, on the other hand, which remains in Euclidean form and which is characteristic of what is called the *orientation* of the metric (the equivalent, in other words, of the orientation group, or position, of the crystal), changes from place to place. All of this becomes even more intellectually exciting when Weyl recalls the famous passage in Riemann's dissertation where he explains that the question of the discrete or continuous nature of space has not been resolved: either the reality on which our space is based is discrete, or it is continuous, and we must look to a force acting from the outside for an explanation of the metric relationships we observe.

To sum up, if I have understood him correctly, Weyl is attempting to establish a series of dichotomies that correspond to one another term for term: nature/orientation, fixed internal structure/variable external agent, discontinuous/continuous. Of course none of this is completely developed, but as always with Weyl, for whom the description "an imaginative and elegant mind" could have been invented, it does give food for thought.

As we have seen, it is no exaggeration to say that the new horizons opening up in crystal science, both in the most basic theoretical research and in technological applications, are turning out to be much more exciting than they seemed at first glance. But the picture would still be incomplete if I did not describe the route by which crystallography has met up with the very latest in astrophysical research.

FROM MINERALOGY TO ASTROPHYSICS

We have seen that for a long time, crystallography remained a close companion of mineralogy; for centuries, it was in fact an integral part of it. Its scientific credibility has occasionally suffered as a result. In an interesting historical twist, having liberated itself from its embarrassing guardian, it now rejoins it under new circumstances. But it is not the "old" mineralogy that is coming back to crystallography; it is a rejuvenated science within what is now called "geophysics." This rejuvenation is the consequence, of course, of new developments in Earth observation methods: for example, remote sensing is being used to find out about the rocks located in certain still inaccessible regions of our planet. But most of all, it is progress in astrophysics that is bringing the two disciplines together. Not just because rocks have been collected on the Moon and are now being carefully scrutinized, or because more and more attention is being focused on meteorites, but also because astrophysics has been reunited with the ancient cosmogonic preoccupations of astronomy. Although scenarios for the origin of the Universe are still the subject of lively debate, a theory of the formation of the solar system appears today to be fairly well-defined, despite a few obscure areas. If the formation of the Earth can be dated to four and a half billion years ago, and if its evolution has proceeded, from the "primitive nebula" stage, in accordance with the cooling process that is postulated, the question of the origin of rocks can be reformulated in a broader context, with more precise instruments than were used previously.

From this point of view, experimental work in current crystallography has the great advantage of offering us an inverse procedure. The idea is not to reconstitute a formative process on the basis of its result (a crystal in a rock unearthed by a mineralogist). Instead, as I have already said, one engages in strictly controlled, directed production of artificial crystals. If the intent is to produce a crystal with a certain specific shape, there are ways of, for example, preventing one face from "growing;" it is possible to produce deliberately selected defects (a capability used to maximum advantage by the electronics industry); and it is therefore also possible to simulate in the laboratory the growth conditions postulated by a particular evolutionary model, and to see whether or not the results correspond to observed reality.

Is it possible, from this evidence, to outline the scenario for a formative process? Certainly not, but at the very least this represents a powerful experimental resource for testing such scenarios once they emerge from the combined work of geophysicists and astrophysicists.

III

FROM LITTLE PHYSICS TO BIG PHYSICS—THE SOCIAL DESTINY OF RESEARCH

CRYSTALS IN BIG SCIENCE

The term "Big Science" has recently come into fashion as a term for the kind of high-energy physics that requires substantial apparatus and investment; "little physics," in contrast, describes research that makes no such demands. It has been acknowledged for some time that research in Big Science is in fact the exclusive province of researchers in the major industrial powers; a rich country's luxury, whose results the others must simply sit and wait for. I should also add that despite appearances (the results in question are, after all, usually published), the industrial countries monopolize the results themselves, always managing to sell what comes out of their own laboratories because they hold commercial rights to the technological novelties that result from them.

But what about "little physics"? Is it more equitably distributed over the planet? From this standpoint, crystallography represents a very revealing instance of a general trend that I consider disastrous. The raw material does not cost very much up to the point where one starts manufacturing crystals, and for a long time the cost of the research equipment remained reasonable. Crystallography was therefore an "affordable" science all over the world. But obviously everything has now changed with the advent of synchrotron radiation. I have described how synchrotron radiation arrived as a by-product of nuclear physics research; today, accelerator tunnels are being specifically earmarked for this purpose, designed so that the radiation emitted by the accelerated particles (generally positrons) is as useful as possible for research and its applications. The characteristics of this radiation—extremely powerful and "white" (that is, not limited to a single wavelength)—make it infinitely superior to the radiation obtained from traditional x-ray sources. All of a sudden, crystallography has been radically transformed: experiments that used to take hours now take just minutes, and new fields of inquiry, unthinkable at a time when only a few wavelengths were available, become possible. But since the technology which allows such marvels is extremely costly, only the developed industrial countries can get on this train; the others must simply stand and watch it pass, despite all the efforts to develop so-called "cooperative" programs.

So the gap between countries where research is done and those where it cannot be done continues to grow, spreading now to a new field. And this is only one example among many. The economic consequences of such a situation are obvious, but we must not underestimate the intellectual and social repercussions: it is the spirit of research that is threatened with extinction for three quarters of humanity. Speeches touting a new "age of science" are at the very least hypocritical when they "forget" this fact. One might object that provisions have been made, in the case which we are discussing, for countries with no accelerators of their own to have access to one, just as astronomers can reserve observation time on a telescope. But in practice, this arrangement can never come out in their favor: any beam utilization project is submitted to a "program committee" made up of researchers working on site (that is, Americans, Europeans, Japanese, or former Soviets). How could they fail to give preference to research being done in their own "community," whose preparation and development they have been monitoring?

But even assuming that a project proposed by researchers from a developing country is accepted (which does happen), bear in mind the difficulties implied in putting it into practice. Since these kinds of experiments generally last several days at a stretch, and once a team's turn is over it has to wait several months for more "beam time," the researchers must arrive on the appointed day with all their preliminary research complete, and string all their experiments together uninterruptedly for the allotted time. The inequality here is flagrant, not just because it is easier to complete preliminary research if one belongs to a laboratory in a major industrialized nation than if one is isolated, for example, in an African country, but also, quite trivially, because research teams have to travel. Now although it is not very expensive to send three or four people from Paris (or even London) to Grenoble, it is a very different matter when you are starting out from Dakar. We have arrived at a point where many countries quite simply cannot finance this kind of travel for their researchers.

The research world is therefore by no means enjoying a phase of happy cooperation; on the contrary, an abyss is opening up dramatically before our very eyes. In more and more areas of research, a growing gap separates those who do research and those who are simply invited to apply what someone else has discovered. That is not all: it is now out of the question for anyone to do crystallography without having the powerful calculation resources represented by computers. And once again, only the major industrialized nations produce computers. Moreover, this is only one specific example of a general situation: all those who do not possess calculation resources, that is, more and more powerful computers that need to be upgraded frequently, are excluded from research and consequently from the results that research delivers. India certainly represents the single notable exception; but will India, whose crystallography community has prospered so far while acquiring its equipment at low cost from Great Britain, be able to retain its standing in this area with the coming of the synchrotron? That is by no means certain.

Consider one of the poorest countries in the world: Madagascar, where I spent some time as a result of scientific cooperation agreements. Here the situation is very clear-cut: it is impossible to do research. The most elementary laboratory equipment, even on a small scale, is not there, from simple adhesive tape to electronic components, which had to be imported with enormous effort from Réunion when it was found that the laboratory "stockroom" was pitifully empty. As far as science is concerned, I can confidently state that the countries hypocritically referred to as "developing" are, on the contrary, set on a course of accelerated under-development, and this disastrous process today appears irreversible.

THE LANGUAGE OF PHYSICS

Educational systems are only making matters worse. No sooner do we begin considering this than we stumble on a problem that is universal in scope. While it is glaringly apparent in higher education, the difficulty is rooted in secondary education, and, much further back, in elementary schooling. There have been plenty of complaints that students never even really master their own native languages. This observation is obviously just as valid for American (or French) students as for African ones, but in the latter case, the difficulty is aggravated by the fact that when they reach a certain level of scientific training, most of these students must go finish their studies in the former colonial mother country or elsewhere in the industrialized world, and must therefore think in a language which is not their own!

You might object that all in all, this question is less dramatically important when science students are involved, and that the obstacle is greater in the case of literary studies or the like. That would be a serious misconception, based on an obfuscatory concept of scientific practice. Learning or doing research does not, in any field—and no more in quantum mechanics than elsewhere—mean applying formulas or manipulating them to fit some naked "reality." Research means thought, and to think inventively, all the subtle resources of language—from which thought cannot arbitrarily be separated—must be mobilized. A person who does not know his or her own language as thoroughly as possible is therefore handicapped as a researcher.

I would like to illustrate this remark with an example borrowed from my experience with students, and extracted from the armamentarium of quantum mechanics, so often called upon earlier in this book. One of its basic statements, as we have seen, is that a system (let us call it an atom) can be in a variety of "states," and that there exist a certain number of these states which are privileged: the inherent states of a physical quantity, privileged in the sense that a measurement of the physical quantity yields a unique result when the system is in one of these inherent states. We then say that it has a specific value of the quantity in question, which is not true when the system is in an arbitrary state. Because they have not mastered the meanings of words as simple as "state" or "value," "be" or "have," students often say that the system "*has* an inherent state" instead of "*is* in an inherent state corresponding to a certain quantity;" or that it "*is* in an inherent value" instead of "*has* an inherent value." I am prepared to bet that if a stock market price were being discussed, these same students would say that a certain stock *has* such-and-such a value; and that in everyday life they do not talk about *having* a state of excitement. But apparently the rules of syntax and semantics are not sufficiently natural and constraining to prevent them from committing barbarisms as soon as the topic turns to physics. This linguistic impediment then inevitably constitutes an obstacle in the mathematical formulation of the problem itself. Specifically, if one does not really believe that the system *is* in a state, there is confusion between state and value, which are two completely different mathematical concepts: one corresponds to a vector in a Hilbert space representing the state *in* which an object can *be*; and the other to a number in the real set, a value which a system can *have* or take on. At first glance, it appears that when a student makes a mistake, he or she has stumbled over the mathematical formulation; in reality, it is a linguistic error which precipitated the mathematical mistake.

A thorough knowledge of a language, in order to be able to think easily, quickly, subtly, and inventively in it, seems to me an absolute necessity for a researcher. Such skills are determined early on, in childhood. However, every ideology to which elementary and then secondary school students are subjected establishes a clear distinction between "literary" and "scientific" studies. Although this distinction is a convenient way of establishing "tracks," a weakness of all educational administrators, it is based on a caricature of scientific activity. Students who opt for research find out later that they have been victimized by this sham that is obligingly maintained in the name of "specialization"—necessary, yes, but having nothing to do with organized illiteracy. These caustic remarks lead to another question. What good are all these efforts, since in any case the language that has now established itself in the research world is not French, or Arabic, or Japanese, but a sort of English? Witness all the conferences and, especially, publications in the international journals that matter, for which English is a requirement.

This almost unrelieved domination leads, especially in France, to some nostalgic recriminations. We have even seen certain ministries trying to stem the flood with laughable administrative actions. Is the phenomenon irreversible? Apparently; however, it is nothing new. At any moment in history, there is no way to prevent an international language from affirming the international nature of science. It was French in the 18th century (although too often we forget that), thanks to two centuries of first-class scientific output; then German in the last century, for the same reasons; and now English. We shall see what happens tomorrow. How should we react to this international language if we are not part of the English-speaking world? Just as we did to German in the 19th century: by learning it thoroughly! To take one example among many, remember that Niels Bohr was Danish, and that he expressed himself in

German and in English. He knew these two languages sufficiently well to choose one or the other depending on what he wanted to say: he wrote certain texts in English, and ended up later translating them into German. We can now read both versions: they are by no means what one would call literal translations, but rather true translations that enlist the respective resources or "genius" of each language.

The threat hanging over the international scientific community is not, in reality, the fact that it is being forced to express itself in English. That complaint is wide of the mark, and only makes the peril worse. The danger is that non-English speakers are making do with merely an approximate knowledge of this language. The remedy is to require readings of English *literature* in every preparatory class for universities and for the major scientific institutes. What is more, the incredible mishmash that is spoken (if one can call it that) at international conferences and written in journals is having an erosive effect on the language of native English speakers themselves, who must communicate only an approximation of ideas which really require the greatest precision.

CONCLUSION

To conclude, let us return to the crystal itself, which in my opinion presents considerable epistemological appeal. What we see is an object that suddenly emerged into the theoretical activities of the physics community, and then, little by little, into the other sciences. It is an object which is now laying down the law of its imperfections, whereas before, when it was still an outsider, it embodied the perfection of a natural order. It might seem at first, therefore, that the object had consumed its theory. But theory instantly reacted, dissociating itself and reorganizing itself around those same imperfections and irregularities that it had to suffer. And here is undoubtedly where the case of the crystal, despite its uniqueness, reveals a general trend in present-day physics. Having looked high and low for regularity, scrutinized the innermost structures of matter, and taken great pains to instill order (if only statistical) in them, physicists are now taking possession of phenomena that they had cast aside: turbulence and chaos, areas in which yesterday the equations looked too complicated, the parameters too numerous. Do we owe this great shift solely to the power of computers which enable us to comprehend all those parameters and solve those innumerable equations at lightning speed? They play their part, to be sure; but there is more to it, and greater depth. What physicists have suddenly realized is that these previously unapproachable problems in fact embodied the same structure as those they had come to know while developing the quantum theory. It was mathematics—in this case group theory—that opened their eyes. Gaston Bachelard called this, if I am not mistaken, the "inductive value of mathematics." It is not a simple analogy, but rather a recognition of the fact that the same mathematics is at work in one branch of physics as in another. The mathematics can be rationally coordinated, gathered under a single "umbrella." As we have seen, nothing would have happened if someone had not realized that group theory was what governed not only the quantum behavior of subatomic matter, but also that macroscopic object called a crystal.

But this story seems to me to convey an even greater lesson: it repudiates a certain very widely held misconception of the history of contemporary physics. In the glare of the intellectual marvel of the "quantum revolution," to use Victor Weisskopf's term, it was believed that a "new world" had been discovered, one completely unknown previously. But although it is correct to state that this world was in fact "new" in the sense that its laws could not be deduced from prior physics, it is incorrect to claim that its effects had never been perceived. After all, quanta existed before they were discovered and given a mathematical formulation! The first crystallographers had really and truly, without knowing it, brought to light some spectacular effects of the atomic (and hence ultimately quantum) structure of matter by observing the disconcerting arrangement of crystal surfaces. They correctly believed that the key to the mystery lay in their configuration on a microscopic scale, although they lacked the experimental and theoretical means to follow up their intuition.

The example is a striking one, but there are more. After all, tables were solid, the sky was blue, the night sky was black, and electricity ran through wires well before the beginning of this century. We know today that only quantum mechanics can account for all these phenomena. But quantum mechanics emerged from crystallography as Napoleon emerged from the young Bonaparte. It is something that has always amazed me, and something that probably is worth further reflection.

BIBLIOGRAPHY

CURIE, P., *Œuvres* [Works],* Paris, Gauthier-Languereau, 1908; Archives contemporaines, 1984.

AUTHIER, A., CURRIEN, H., KERN, R., and ROUSSEAU, M., "Cristaux" [Crystals],* in *Encyclopedia universalis.*

GUINIER, A., *La Structure de la matière* [The structure of matter],* Paris, Hachette, CNRS, 1980.

HAÜY, R., *Traité de cristallographie* [Treatise on crystallography],* Paris, 1822; Bruxelles, Culture et civilisation, 1968.

LAUE, M. von, *Röntgenstrahl Interferenzen* [X-ray interference],* Leipzig, 1941.

PASTEUR, L., VAN'T HOFF, J., and WERNER, A., *Sur la dissymétrie moléculaire* [On molecular dyssymmetry],* preface by J. Jacques, postscript by Cl. Salomon-Bayet, Paris, Ch. Bourgeois, "Epistémé" series, 1986.

SHUBNIKOV, A. V., and KOPTISK, V. A., *Symmetry in Science and Art*, translated from the Russian, New York, Plenum Press, 1974.

WEYL, H., *Symmetry*, Princeton, PUP, 1952.

WEYL, H., *Philosophy of Mathematics and Natural Science*, Princeton, PUP, 1949.

WEYL, H., *Gruppentheorie and Quantenmechanik* [Group theory and quantum mechanics],* Leipzig, 1931.

WIGNER, E., *Symmetries and Reflections* (*Scientific Essays*), Cambridge, MIT Press, 1967.

* These references have not been published in English.